What Do You See in the Moon?

by Lara Winegar

People around the world see different pictures in the moon.

Do you see a face?

Some people see the Man in the Moon!

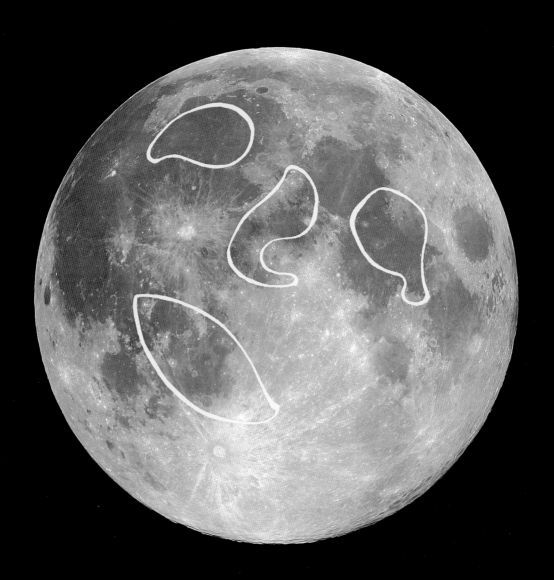

Some people in Germany also see a man in the moon. This man has sticks on his back.

Can you see the man?

Some people in India see a rabbit on the moon.
They tell a story about the rabbit. A rabbit tried
to help a man. The man put the rabbit on the
moon as a reward.

Can you see the rabbit?

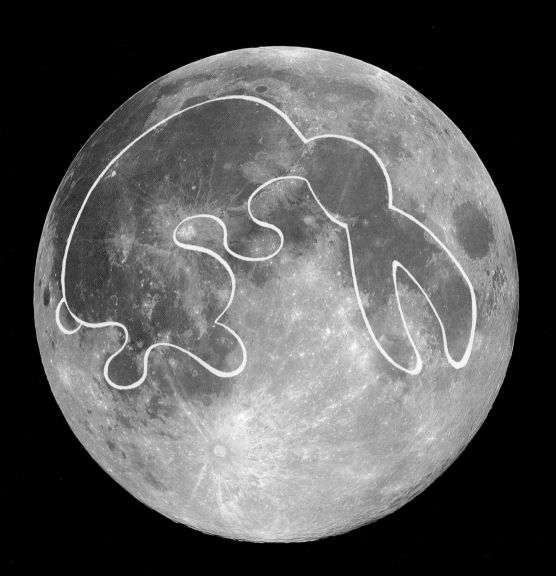

Some Native American people see a toad in the moon. They tell a story about the toad. A wolf tried to catch a toad. But the toad jumped to the moon!

Can you see the toad?

Some people see a lady in the moon. She has a necklace with a jewel in it. The jewel is a crater. A crater is a dent on the moon.

Can you see the lady?

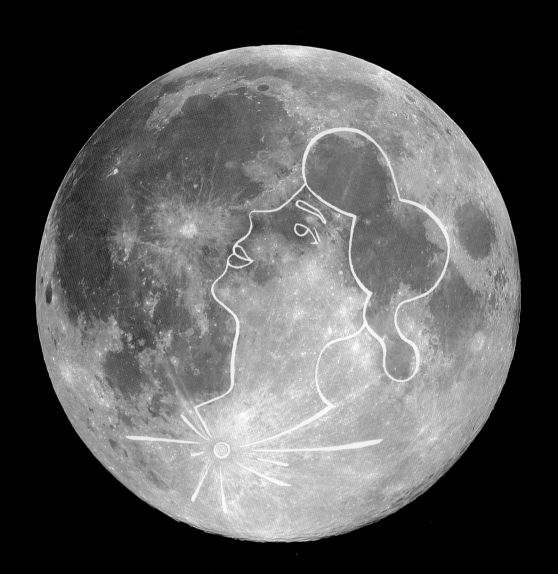

What do you see in the moon?